YOUR KNOWLEDGE HAS VALUE

- We will publish your bachelor's and
  master's thesis, essays and papers

- Your own eBook and book -
  sold worldwide in all relevant shops

- Earn money with each sale

Upload your text at www.GRIN.com
and publish for free

Amalia Aventurin

Multi-Sensor-Core-Logger (MSCL)

GRIN Verlag

Bibliografische Information der Deutschen Nationalbibliothek:

Die Deutsche Bibliothek verzeichnet diese Publikation in der Deutschen National-
bibliografie; detaillierte bibliografische Daten sind im Internet über http://dnb.d-
nb.de/ abrufbar.

Imprint:

Copyright © 2013 GRIN Verlag GmbH
Druck und Bindung: Books on Demand GmbH, Norderstedt Germany
ISBN: 978-3-656-64485-9

Multi-Sensor-Core-Logger (MSCL)

I. Introduction: MSCL Principle

The MSCL-experiment encloses the stepwise measurement of three different parameters: Gamma density, P-wave-velocity (compressional wave travel time) and magnetic susceptibility. Each is measured by different sensors. A photo of the apparatus is shown in figure 1.

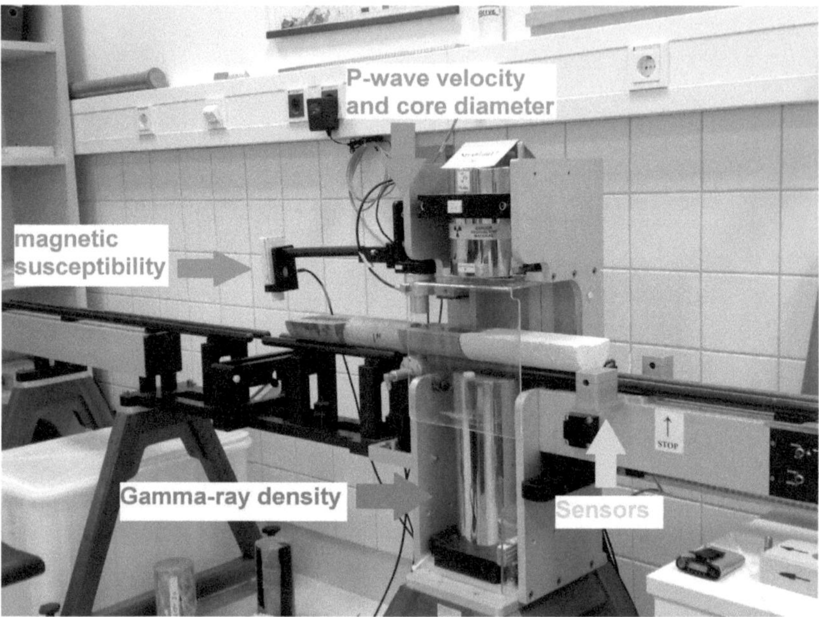

Fig. 1: Used MSCL apparatus with description where the sensors lie.

The four core samples G1, a black stone, coarse-grained and compacted with small mica particles and bigger white quartz inclusions, could be a gabbro and G2 a greenish sandstone with small particles and lesser compaction, each unsaturated and saturated with water are halved and "transported on a stepper motor-driven tracking system" to the sensors. If the rock sample is heterogeneous and the halves don't accord in their mineral composition, you will have now a potential error source. The samples are laid on the tracking system. A motor pushes them first to a laser, where the length is measured, than to the gamma source and then to the P-wave-velocity-sensor. Here you have a second potential error source: When the P-wave-velocity-sensor presses the samples down for measuring, they were lift on the other side. To avoid the lifting the rock samples have to be pressed and so the measurements are not really accurate. At least the samples were driven to the sensor, which measures the magnetic susceptibility. Due to the fact that the samples are a little bit short, we need to put dummies between the samples, which are not measured.

The measurement of the magnetic susceptibility is used to determine the degree of magnetization. The magnetization only depends on the composition of the rock and matrix properties. It is independent from any fluid, which may fill the pore spaces.

The measurement for gamma density bases on caesium^{137} compton-scattering: The denser a sample, the lesser electrons go through the sample. With the electron-intensity, the computer determines at the end the density of the rock sample. The P-wave-velocity bases nearly on the same principle, but here you have a sound-wave (P-wave), which goes through the sample.

Before the measurement can start the apparatus has to be calibrated. Those calibrations have to be done to determine the errors depending on ray energy reduction in consequence of the interaction of the rays with oxygen atoms and on the size of the sample.

The first calibration we have done is the P-wave velocity (V_p). For this we have four reference blocks with a defined thickness, which are placed one after another under the sensor. The values for the reference blocks are shown in table 1. The plotted values are shown in figure 2a. The values for the gate are given for each reference block and where entered manually in the computer.

What has to be considered in this case is that the porosity value for saturated material will be overrated by about 11%, so it has to be corrected by the following equation:

$$\rho_{corrected} = \frac{(\rho_{sat} - \rho_{dry} + \rho_{air}) \cdot 11}{100} \tag{1}$$

	Thickness [cm]	P-wave Travel time [ ]	Gate [µs]
1	1.996	17.55	10
2	3.997	20.60	15
3	5.995	23.75	100
4	8.002	26.55	100

Tab. 1: Values for the four reference blocks. Just the values for P-wave Travel time were measured by the sensor.

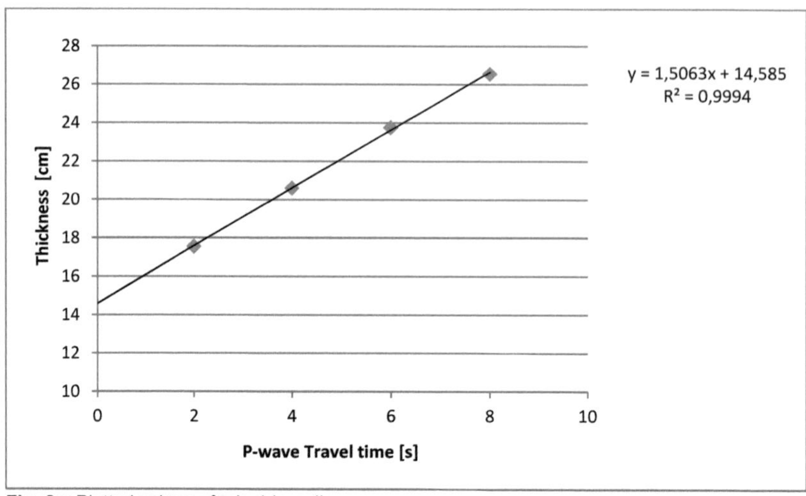

Fig. 2a: Plotted values of tab. 1 in a diagram.

Out of these results a regression line was calculated by an excel sheet:

$$y = 1.5063 + 14.585 \quad R^2 = 0.9994$$

The number 14.585 gives us now the runtime, which is needed for the measurement.

Each measurement step was done every centimeter for 10 seconds for gamma-ray and density and 1 second for magnetic susceptibility.

The next calibration is now done for the gamma-ray density. This is done by 14 aluminum discs which are measured every time for 60 seconds. For this measurement the discs are laid one upon the other to increase mass, but the first measurement is done without any disc – to get a raw value. The results are shown in table 2 and figure 2b.

	Gamma-ray density [cps]		Gamma-ray density [cps]
L_0	23912	L_{1-8}	5721
L_1	20332	L_{1-9}	4714
L_{1-2}	17200	L_{1-10}	3894
L_{1-3}	14440	L_{1-11}	3209
L_{1-4}	12079	L_{1-12}	2645
L_{1-5}	10047	L_{1-13}	2192
L_{1-6}	8332	L_{1-14}	1792
L_{1-7}	6915		

Tab. 2a: Results for Gamma-ray-density intensity calibration

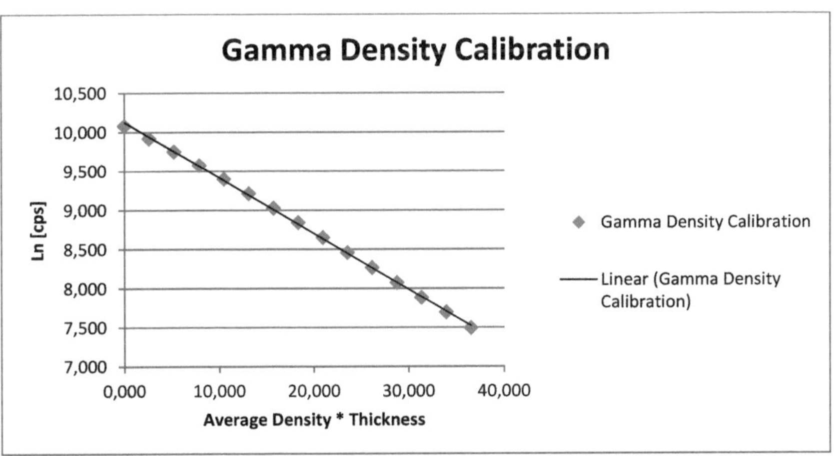

Fig. 2b: Plotted results for Gamma-ray density (Ln [cps]). On the x-axes are the average density·thickness plotted.

The regression line for this polynomial is

$$y = -0.0002x^2 + 0.00652x + 10.093 \quad R^2 = 1$$

and is now used for the calibration. The magnetic doesn't need a calibration, so the sample measurements can start.

II. Experimental measurements and Results

The measured values for two sample halves G1 and G2, dry and saturated are shown in table 3 and 4 and figure 3a, b, c, d.

	P-wave Amplitude	P-wave velocity [m/s]	Bulk Density [m/cm³]
G1 dry	94	-	-
	96	6257.259	2.1278
	94	6330.131	2.1286
	94	6306.577	2.1544
	92	6153.093	2.1341
	93	6139.287	2.1403
	97	6222.816	2.1348
	97	6240.057	2.1261
	91	6229.779	2.124
	79	6129.349	2.1386
	97	6867.946	2.2336
	97	6756.02	2.2137
	97	6910.322	2.2323
	0	4843.845	1.9473
G2 dry	0	4352.326	1.0262
	96	4630.627	1.9036
	97	4699.638	1.9036
	97	4739.642	1.9078
	97	4778.098	1.9241
	97	4812.61	1.9216
	97	4748.36	1.9105
	97	4735.928	1.9127
	97	4755.717	1.9183
	97	4755.717	1.9262
	97	4739.445	1.916
	97	4725.61	1.9067
	97	4745.387	1.9121
	97	4676.665	1.9047
	97	4692.371	1.9084
	97	4722.65	1.9136
	97	5185.599	1.9742
	97	4980.302	1.9693
	97	5037.723	1.9929
	0	3386.809	1.8619

Tab. 3: Measured values for sample G1 dry and G2 dry. Just the values relevant for the calculation are shown.

	P-wave Amplitude	P-wave velocity [m/s]	Bulk Density [g/m³]
G1 **saturated**	97	5408.502	2.0054
	0	3246.051	2.053
	97	6535.233	2.1844
	97	6603.063	2.1703
	97	6767.213	2.2084
	97	6484.685	2.1712
	97	6478.57	2.1655
	97	6466.341	2.1408
	97	6485.109	2.1537
	97	6491.717	2.149
	97	6686.584	2.1918
	97	7294.663	2.2386
	97	7225.949	2.224
	66	6254.233	2.1833
	0	3619.304	2.0121
G2 **saturated**	21	14290.86	1.8724
	97	4814,.86	1.9916
	97	4844.132	2.0157
	97	4837.69	2.0117
	97	4778.493	2.0111
	97	4726.375	2.0097
	97	4858.3	1.9926
	97	4704.376	2.0062
	97	4593.876	1.995
	97	4529.594	1.9784
	97	4598.288	2.0023
	97	4666.219	1.9782
	97	4682.546	1.9788
	97	4809.868	1.9788
	97	4676.423	1.9946
	97	4773.588	2.0159
	97	5150.688	2.0802
	97	5279.492	2.0836
	97	5228.241	2.0801
	97	5072.229	1.5481

Tab. 5: Measured values for G1 saturated and G2 saturated. Just the values relevant for the calculations are shown.

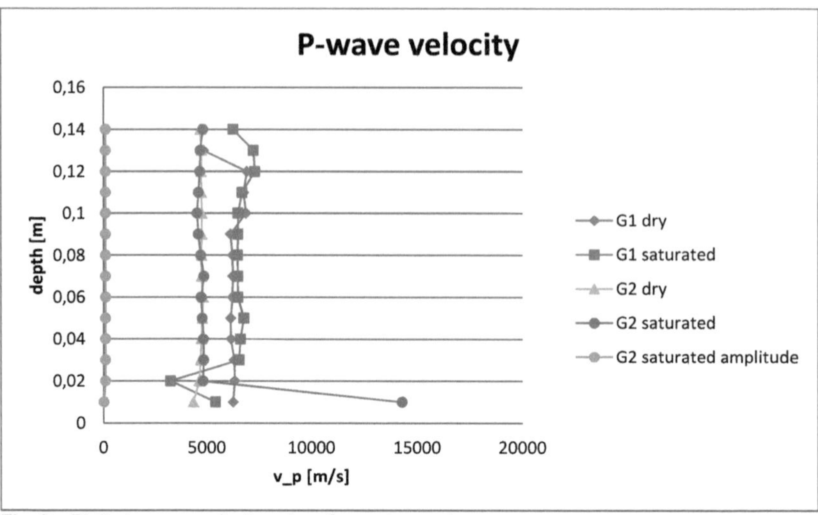

Fig. 3a: Plotted p-wave velocity values for each sample.

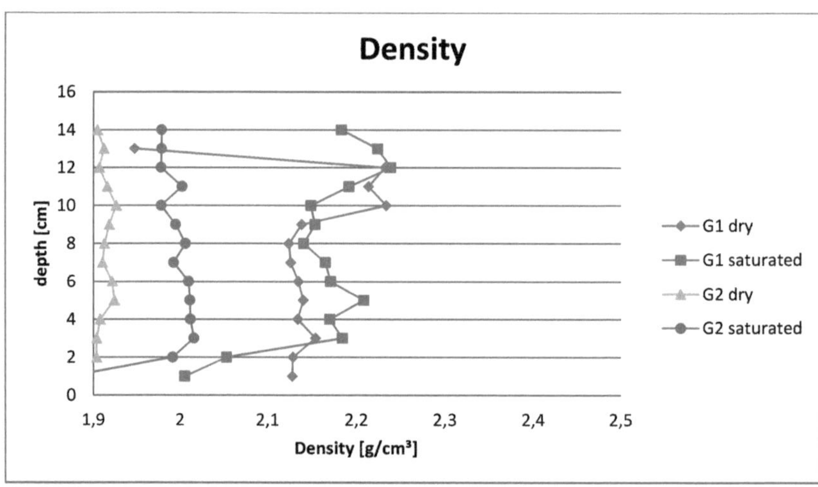

Fig. 3b: Plotted density values for each sample.

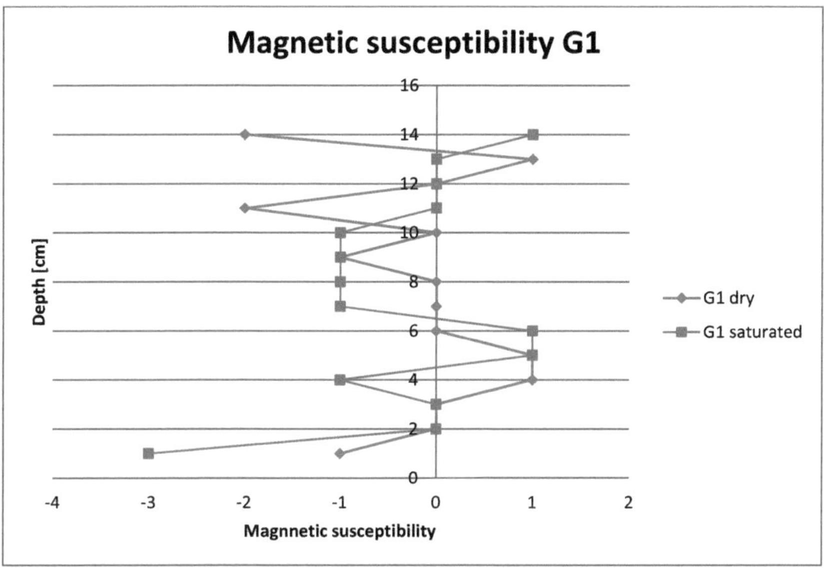

Fig. 3c: Plotted magnetic susceptibility values for sample G1.

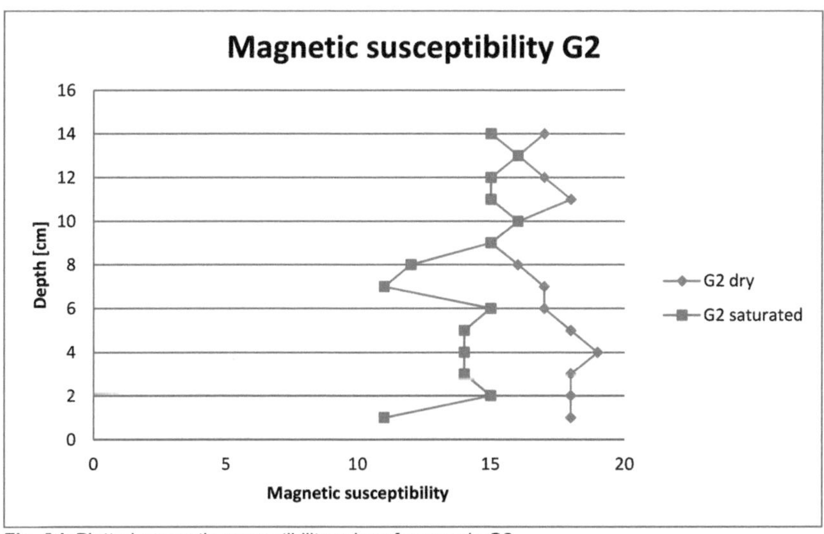

Fig. 3d: Plotted magnetic susceptibility values for sample G2.

The porosity [%] was measured using ρ_{fluid} = 0.9982 g/cm³ and ρ_{air} = 0.0012 g/cm³ (at 15°C) within the following equation:

$$\phi = \frac{\rho_{dry} - \rho_{sat}}{\rho_{air} - \rho_{water}} \qquad (2)$$

The derivation of formula (3) is as followed:

$$\rho_{bulk} = \phi \cdot \rho_{porefluid} + (1 - \phi) \cdot \rho_{matrix} \qquad (3)$$

This base on the gamma-ray density measurement of the dry and saturated core halves.

For both samples the porosity has to be calculated and then equalized to one formula:

$$\frac{\rho_{bulk,dry} - \phi \cdot \rho_{air}}{1 - \phi} = \frac{\rho_{bulk,wet} - \phi \cdot \rho_{water}}{1 - \phi} \qquad (4)$$

Than this formula as solved for ϕ and the results is equation (3)

Where ρ_{dry} is the density of the dried sample and ρ_{sat} the density of the saturated sample.

For this calculation the overrating of 11% for the saturated sample has to be considered.

The Young's modulus [MPa] is measured by the following equation:

$$E = \frac{\sigma}{\epsilon} \quad \text{and} \quad M = \frac{\sigma}{\epsilon} = \rho \cdot v_{\rho}^2 \qquad (5)$$

With this calculation the average results for Young's modulus (5) are:

      G1 dry = 70684.31 Mpa = 70.68 GPa

      G2 dry = 34809.1 MPa = 34.81 GPa

      G1 saturated = 70399.72 MPa = 70.4 GPa

      G2 saturated = 52582.01 MPa = 52.58 GPa

The higher the value for the E-modulus, the greater is the resistivity of the sample against deformation. It characterizes the correlation between elongation and stress for the deformation of solid material.

The mean values for Young's modulus are now taken to calculate the force to compress the samples along their axes by 2cm. To calculate this, the E-Modulus has to be transformed and set into context with density and V_p velocity as shown in the following:

$$V_p = \sqrt{\frac{M}{\rho}} = \sqrt{\frac{1-v}{(1+v)\cdot(1-2v)}} \cdot \sqrt{\frac{E}{\rho}} \rightarrow v_p^2 = \frac{(1-v)}{(1+v)\cdot(1-2v)} \cdot \frac{E}{\rho} \rightarrow E = \frac{\rho \cdot v_p^2 \cdot (1+v)\cdot(1-2v)}{(1-v)} \left[\frac{N}{m^2}\right]$$

$$(6)$$

The force is calculated with the following equation:

$$F = \frac{A \cdot E \cdot \Delta L}{L} \, [N] \tag{7}$$

III. Calculation

III.1 Sample G1

By using the equation (3) we get the following results shown in figure 4.

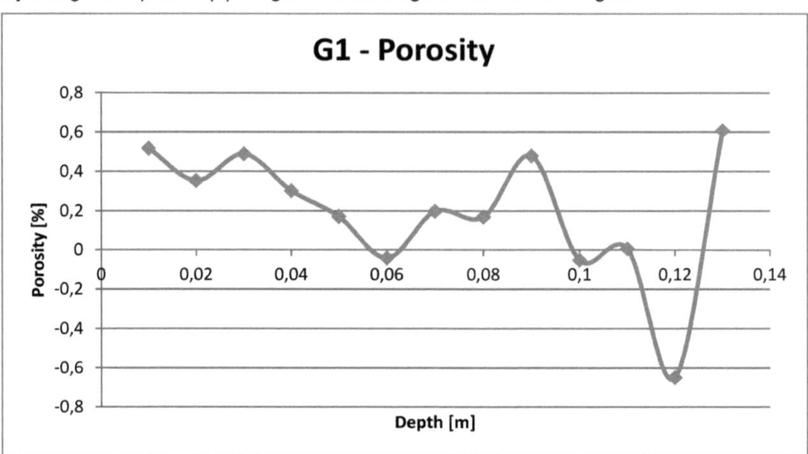

Fig. 4: Porosity Results for sample G1. The negative values are negligible.

The average porosity for sample G1 is 0.12 %. The negative values can be caused by false density measurements. The porosity values for this sample vary a lot, but are always beneath 1%. This leads to the point, that this is a very compacted and heterogeneous rock. The negative values for the porosity are negligible but lead to the argument that at this point the porosity is 0%.

Compared to the given porosity value from the Archimedes' principle of 0.08% the measured range is beneath 0.6%, but this has no significance.

The results for Young's modulus, using the equation (5), are shown in figure 5. The average E-modulus for this sample is $G1_{dry}$ = 84.82 GPa and $G1_{saturated}$ = 84.48 GPa. What can be observed in the picture is that Young's modulus is slightly higher for the saturated sample than for the dried one, this is due to the fact that water is an incompressible medium. The E-modulus depends on lithology and including minerals, etc. Especially for the saturated sample the values vary a lot. This leads to the assumption that this sample includes minerals which are build up on water so that the values rises. In fact the values for the E-modulus for this sample are relatively high, so the rock has a strong resistivity against deformation. Under mechanical stress this rock sample would break the easiest way on the point with the lowest Young's modulus, but even this point is around 80 GPa so you need even here a high mechanical stress.

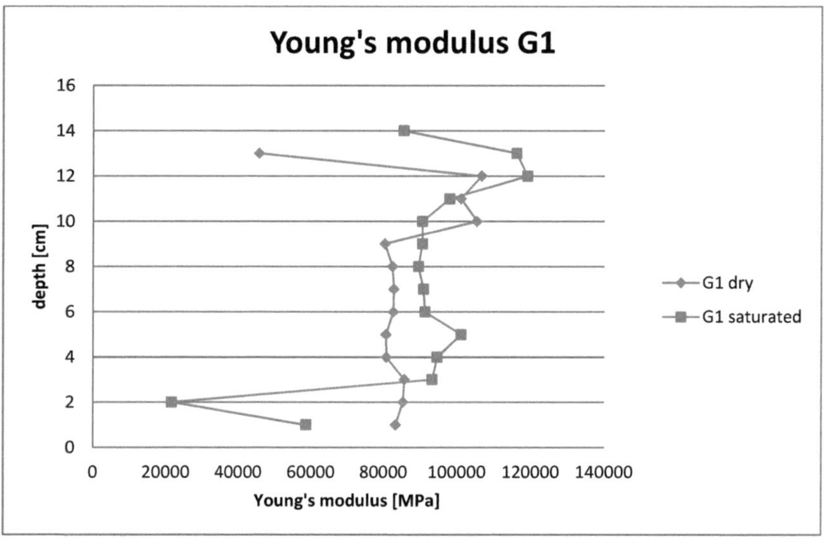

Fig. 5: Results for Young's modulus (E-modulus) for sample G1 dry and saturated.

The mean values for Young's modulus are now taken to calculate the force to compress the samples along their axes by 2cm. With the equations (6) and (7) a diagram, seen in fig. 5a can be plotted.

The mean values are $G1_{dry}$ = 0.01 m² and $G1_{saturated}$ = 0.01 m² for the area, $G1_{dry}$ = 76.73 MPa and $G1_{saturated}$ = 64.91 MPa for the force. Out of the mean values it can be said that the sample would react elastic to deformation, if you would perform a force at 2cm along the axes of the sample. If you perform a force on the whole sample it would react with brittle failure.

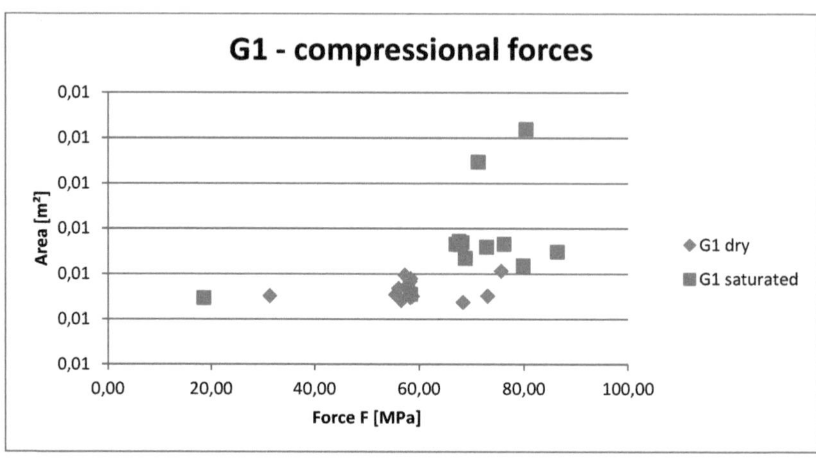

Fig. 5a: The compression forces for G1 plotted against the area.

III.2 Sample G2

In general, G2 shows increased values in p-wave velocity. By using the equation (3) we get the following results shown in figure 6.

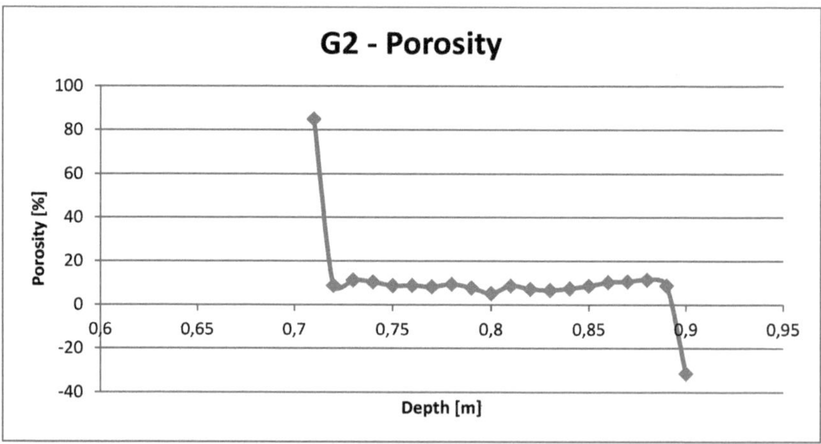

Fig. 6: Porosity Results for sample G2. The negative values are negligible, caused by measurement errors in the beginning and at the end of the sample (which is already mentioned in the Introduction).

The average porosity for sample G2 is 10.59 %. The negative values are caused by measurement errors in the beginning of the sample measurement and at the end and are therefore negligible. The porosity is nearly constant and relatively high, what can fit to the argument that this is a homogeneous sandstone.

Compared to the given value from the Archimedes' method of 12%, the measured values from the MSCL experiment range between 8-15%. So the Archimedes' value is nearly comparable to the MSCL.

The results for Young's modulus, using the equation (5), are shown in figure 7. The values for the E-modulus varies in this sample and are slightly higher for the saturated sample, which can be lead to the argument, that there are water building minerals in the rock. This rock sample has a lower E-modulus, so it reacts easier to deformation with breakage, but for the reason that this is a real homogeneous rock you even need here a high deformation force to break this rock. The lowest point is by about 40 GPa so here could a breakage or deformation happen.

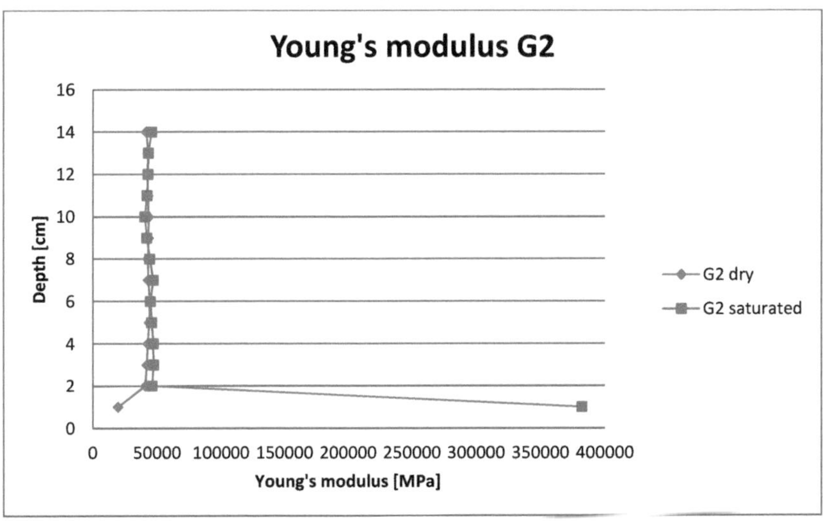

Fig. 7: Results for Young's modulus (E-modulus) for sample G2 dry and saturated.

The average E-modulus for this sample is $G2_{dry}$ = 41.77 GPa and $G1_{saturated}$ = 63.1 GPa.

The mean values for Young's modulus are now taken to calculate the force to compress the samples along their axes by 2cm. With the equations (6) and (7) a diagram, seen in figure 7a can be plotted.

The mean values are $G2_{dry}$ = 0.01 m² and $G2_{saturated}$ = 0.01 m² for the area, $G2_{dry}$ = 21.82 MPa and $G2_{saturated}$ = 30.48 MPa for the force. Out of the mean values it can be said that the sample would react elastic to deformation, if you would perform a force at 2cm along the axes of the sample. If you perform a force on the whole sample it would react with brittle failure.

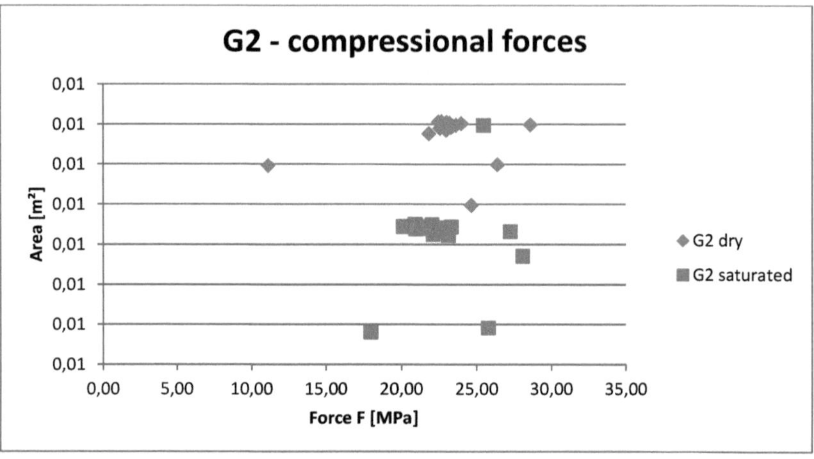

Fig. 7a: The compression forces for G1 plotted against the area.

IV. Error calculation

The following equations are used to the determine the error for the compressional force (8) and the E-Modulus (9):

$$\Delta F(A,E) = \sqrt{(\tfrac{E*dl}{L} * \Delta A)^2 + (\tfrac{A*dl}{L} * \Delta E)^2}$$

(8)

$$\Delta E(\rho,Vp) = \sqrt{(Vp^2 * \tfrac{(1+v)*(1-2v)}{(1-v)} * \Delta\rho\,)^2 + (\rho * 2Vp * \tfrac{(1+v)*(1-2v)}{(1-v)} * \Delta Vp\,)^2}$$

(9)

Where A is the area of the cross section [m²], E is the E-Modulus of the sample [GPa], ρ is the density of the sample [kg/m³] and V_p is the p-wave velocity in m/s.

The results for the error calculation for Young's modulus and the compressional force are shown in table 5.

	Δ E-Modulus [GPa]	Δ F [GPa]
G1$_{dry}$	11.58	0.11
G1$_{saturated}$	26.002	0.026
G2$_{dry}$	6.3	0.006
G2$_{saturated}$	37.29	0.034

Tab. 5: Errors calculation after Gauß for Young's modulus and the compressional force for each samples.

The standard deviation for the force and Young's modulus are listed in table 6.

	Standard Deviation	
	E-Modulus [MPa]	F [GPa
G1$_{dry}$	12.64	0.011
G1$_{saturated}$	23.39	0.026
G2$_{dry}$	6.62	0.006
G2$_{saturated}$	62.77	0.034

Tab. 6: Standard deviation for Young's modulus and the compressional force for each sample

V. Interpretation and Conclusion

The values for p-wave velocity and bulk density don't vary a lot for the dry and saturated sample G2. For the dried sample the measured values are slightly smaller than for the saturated sample. This is caused by including water, which causes an increase in travel time. For sample G1 great variations in density can be observed. The scale of variation can be caused by lithological inhomogeneity or by wrong thickness measurements.

The magnetic susceptibility does not correspond to the porosity but can vary due to the inhomogeneity of a rock and including therefore more or less minerals in one of the halves. What can be observed is, that for sample G2 the magnetic susceptibility is higher than for G1. This can be affected by minerals, which have a magnetic sensitive reaction, e.g. magnetite or fayalite.

The porosity values nearly fit to the given values from the Archimedes' measurement, but are slightly smaller. The results getting from the MSCL experiment lead to the assumption that sample G1 is a densely packed gabbro and G2 is a porous sandstone in comparison to literature values.

Just by given values for Young's modulus it is nearly impossible to give a rocks' name, because there are other factors such as voids, type of mineral, layering or packing which effects the rock locally.

The elastic and inelastic deformation depends on stress, temperature, strain, depth, etc. If there would happen an elastic or inelastic deformation or even a breakage can't be said for the whole sample. Out of the results for Young's modulus you could say, that sample G2 would easier deform than G1 because of the higher values for the E-modulus. To say in detail when, how and why needs further research, e.g. on a "rock compactor" ("Gesteinspresse"), which would exceed the report.

The measurement with the MSCL tool has many advantages such as measured values are distributed laterally, heterogeneities are detectable, therefore profile sections are possible and the measurement in the lab is a good equivalent to geophysical borehole logging. This method is just advisable for samples with few inhomogenities, because the samples are split into halves and you can't be sure that every halve is 100% equal to the other.

VI. References

- http://www.marum.de/Mutli-sensor_core_logger_MSCL_for_non-destructive_analyses_of_physical_properties.html (28.12.2012)

- http://web.ead.anl.gov/resrad/datacoll/porosity.htm (29.01.13)

- Mavko, G., Mukerji, T. and Dvorkin, J. (2009) The Rock Physics Handbook: Tools for Seismic

- Analysis of Porous Media. Cambridge University Press, 2nd edition.

- Parasnis, D. (1997) Principles of applied geophysics. Chapman Hall, 5th edition.

- Turcotte, D. L. and Gerald, S. (2002) Geodynamocs. Cambridge University Press, 2nd edition.